DECODING THE INSTALLATION OF SOLAR PANELS

1ST EDITION

How to create and calculate your photovoltaic systems for any application

Karl Franklin, Delfin

© Copyright 2018 – Alán Adrián Delfín Cota. **All rights reserved.**

The contents of this book may not be reproduced, duplicated or transmitted without direct written permission from the author.

Under no circumstances will any legal responsibility or blame be held against the publisher for any reparation, damages, or monetary loss due to the information herein, either directly or indirectly.

Legal Notice:

This book is copyright protected. This is only for personal use. You cannot amend, distribute, sell, use, quote or paraphrase any part of the content within this book without the consent of the author.

Disclaimer Notice:

Please note the information contained within this document is for educational and entertainment purposes only. Every attempt has been made to provide accurate, up to date and complete, reliable information. No warranties of any kind are expressed or implied. Readers acknowledge that the author is not engaging in the rendering of legal, financial, medical or professional advice. The content of this book has been derived from various sources. Please consult a licensed professional before attempting any techniques outlined in this book.

By reading this document, the reader agrees that under no circumstances is the author responsible for any losses, direct or indirect, which are incurred as a result of the use of information contained within this document, including, but not limited to, —errors, omissions, or inaccuracies.

Dedicatory

Dedicated to my enemies, that had helped me grow in my career

Table of Contents

INTRODUCTION 7
CHAPTER ONE 9
  SOLAR POWER 9
  TYPES OF SOLAR ENERGY 10
  CHAPTER TWO 19
  THE IMPORTANCE OF SOLAR ENERGY 19
  THE FUTURE OF SOLAR ENERGY 23
CHAPTER THREE 32
  THE MOST EFFECTIVE METHOD TO UNDERSTAND ELECTRICITY: WATTS, AMPS, VOLTS, AND OHMS 32
CHAPTER FOUR 58
  WHAT IS A GRID-TIED SYSTEM? 58
  OFF GRID SOLAR SYSTEM 62
CHAPTER FIVE 71
  PHOTOVOLTAIC SOLAR 71
  HYBRID SOLAR SYSTEM 75
CHAPTER SIX 86
  FACTS ABOUT SOLAR ENERGY 86

TIPS FOR SOLAR SHOPPERS 93
FACTORS THAT AFFECT SOLAR PV SYSTEM
EFFICIENCY 96
CHAPTER SEVEN 109
SOLAR SYSTEM SIZING 109
CONCLUSION 128

INTRODUCTION

Solar energy is an energy that originates from the sun. Consistently the sun emanates or conveys a broad measure of energy. The sun radiates more energy in one moment than individuals have utilized since the very beginning!

Where does this energy originate from? It arises from inside the sun itself. Like different stars, the sun is a significant gas ball made up generally of hydrogen and helium. The sun creates energy in its center in a procedure called nuclear fusion. Amid nuclear fusion, the sun's incredibly high weight and hot temperature influence hydrogen molecules to break into pieces and their cores (the focal centers of the particles) to wire or consolidate — four hydrogen cores breaker to end up one helium iota. In any case, the helium particle weighs not precisely the four cores that solidified to shape it. Some issue is lost amid nuclear fusion. The lost problem is discharged into space as brilliant energy.

It takes a vast number of years for the energy in the sun's center to advance toward the solar surface, and after that only barely eight minutes to venture to every part of the 93 million miles to earth. The solar energy goes to the earth at a speed of 186,000 miles for each second, the speed of light.

Just a little segment of the energy transmitted by the sun into space strikes the earth, one section in two billion. However, this measure of energy is gigantic. Consistently enough energy hits the United States to supply the country's energy requirements for one and a half years!

Where does this energy go? Around 15 percent of the sun's energy that hits the earth is reflected again into space. Another 30 percent is utilized to vanish water, which, lifted into the environment, deliver's rain-fall. Solar energy additionally is consumed by plants, the land, and the seas. The rest could be utilized to supply our energy needs.

CHAPTER ONE

SOLAR POWER

Solar power is created by gathering sunlight and changing over it into electricity. This is finished by utilizing solar boards, which are expansive level boards comprised of numerous individual solar cells. It is regularly utilized in remote areas, in spite of the fact that it is ending up increasingly famous in urban zones also.

TYPES OF SOLAR ENERGY

Solar energy is an important decision for a lot of customers, organizations, and associations that are looking to create greener energy just as save money on their energy bills. In reality, the more significant part of us know about the look of photovoltaic boards and properly see solar energy as energy which is created by the sunbeams that achieve photovoltaic boards.

Be that as it may, there is substantially more to solar energy both regarding set-up just as the kinds of solar energy. This book will reveal some insight into the diverse sorts of solar energy with the goal that our per users can settle on better-educated decisions while picking some solar energy that they might want to see introduced in their homes.

Solar energy innovation depends on the capacity to change over the sun's light into usable energy. Be that as it may, it can do as such in an assortment of routes to give heat, light, boiling water, electricity, and notwithstanding cooling to houses, structures or even modern buildings.

1. Photovoltaic systems

A standout amongst the most well-known approaches to utilize solar power is to utilize photovoltaic systems or as they are likewise known solar cell systems, which create electricity straightforwardly from sunlight.

The essential guideline behind this innovation is like what we find in clock or number crunchers that are powered by the sun!

The semiconductor materials utilized in these solar energy systems retain sunlight which makes a response that creates electricity – to be correct, the solar energy thumps the electrons free from their iota which makes them move through the semiconductor material and deliver energy.

Today, solar board innovation can assimilate and change over into energy the more significant part of the unmistakable light range and about a portion of the bright and infrared light range.

Solar cells are generally consolidated into modules that hold around 40 cells, and all in all can make the grade regarding a few meters as an afterthought. In light of their portable size and offer, these flat plate photovoltaic exhibits can be mounted at a settled edge confronting south, or they can be installed on a GPS beacon that pursues the sun, enabling them to catch the most sunlight through the span of a day.

A few of these photovoltaic exhibits would be expected to give enough power to a family; however for an expansive electric utility or mechanical application, many clusters would be required and these eventual interconnected to shape a separate, substantial photovoltaic system.

2. Thin film solar cells

Likewise, this sort of innovation can similarly be kept running with thin film solar cells which use layers of semiconductor materials just a couple of micrometers thick. This has made it workable for solar cells to twofold as housetop shingles, rooftop tiles, building veneers, or the coating for sky facing windows or atria expanding utilization of the accessible space from where sunlight would be caught.

3. Fascinating reality:

There have been real enhancements in the proficiency of this innovation especially as far as catching and changing over sunlight. For instance, the primary solar cells worked during the 1950s, had efficiencies of fewer than 4%. Today's innovation, by and large, offers capabilities of around 15 %.

4. Solar water heating systems

The second sort of solar energy is solar high temp water which as the name proposes includes the heating up of water utilizing the sun's heat. The thought behind this comes straight from nature: the shallow water of a lake or the water on the shallow end of shoreline is generally hotter contrasted with more deep water. This is because the sunlight can heat the base of the lake or seashore in the shallow regions, which thus, heats the water. Along these lines, a system has been created to mirror this: solar water heating systems for structures are comprised of two sections, the solar authority and a storage tank.

The most well-known authority is known as a level plate gatherer who is mounted on the rooftop and appearances the sun. Little cylinders go through the case and convey the liquid – either water or other liquid, for example, a radiator fluid arrangement – to be heated. As heat develops in the gatherer, it heats the liquid going through the cylinders. The storage tank at that point holds the hot fluid.

5. Solar power plants

A third way we can bridle the sun's power for energy is solar electricity; this is generally utilized in modern applications. As the vast majority of us know, most power plants use non-sustainable petroleum derivatives to boil water.

The steam from the boiling water influences a vast turbine to pivot which thus enacts the generator to create electricity. Along these lines of generating electricity is terrible for both the earth and our wellbeing is given the discharge of ozone-harming substances and air poisons from the copying of petroleum derivatives.

In any case, fortunately, another age of power plants is being presented which depend on solar power! These plants utilize the sun as a heat source, and they can do as such in three diverse ways:

Allegorical trough systems catch the sun's energy through long rectangular, bent mirrors that are tilted toward the sun. Along these lines, they help center sunlight around a pipe that contains oil. The oil is heated and utilized at that point used to boil water in a conventional steam generator to deliver electricity.

A dish/motor system utilizes a reflected dish looking like fit as a fiddle a substantial satellite dish which gathers and focuses the sun's heat onto a recipient. This collector retains the heat and exchanges it to the liquid inside a motor. The heat makes the liquid extend against a cylinder or turbine and produces mechanical power. This power is utilized to run a generator or alternator to deliver electricity.

A power tower system utilizes an important field of mirrors to think sunlight onto the highest point of a pinnacle, where a recipient containing liquid salt sits. The salt's heat is utilized to produce electricity through a conventional steam generator. Molten salt holds heat effectively, so it tends to be put away for quite a long time before being changed over into electricity. That implies power can be delivered on overcast days or even a few hours after sunset.

6. Passive solar heating

A further way that solar power can be tackled is through the technique for passive solar heating and day lighting. This is anything but another idea – to be sure, old human advancements, for example, the Anasazi Indians in Colorado had created passive solar structure in their abode.

The effect of the sun is straightforward: advance outside on a warm sunny day, and you can feel the sun. With the legal structure, structures can likewise "feel" the sun's energy.

For instance, south-bound windows will get more sunlight while structures can likewise fuse materials, for example, sunlit floors and dividers that assimilate and store the sun's heat.

These materials heat up amid the day and gradually discharge the heat around evening time when heat is generally required. Other plan highlights, for example, a sunspace, which looks like nurseries, focus a great deal of warmth which with the correct ventilation can be utilized to heat a whole building [5]. Such highlights boost the immediate increases from the sun's heat yet also sunlight itself. The surprisingly better news is that on unusually hot days, there are approaches to guarantee these highlights don't overheat structures.

CHAPTER TWO

THE IMPORTANCE OF SOLAR ENERGY

A standout amongst the most alarming issues of today is the increasing cost of energy. Energy costs are on the ascent as Earth's resources are being drained gradually. Fortunately, innovation has given new resources from common substances, for example, solar energy. Even though interest for energy keeps on ascending, there are things each mortgage holder can do to bring down their costs and help the earth. On the off chance that you are searching for ways to join the advantages of solar energy into your home, converse with your renewable energy company and use these contemplations to enable you to settle on a choice.

✓ SOLAR ENERGY HELPS THE ENVIRONMENT

Solar energy is one of those renewable resources that is extraordinary for the earth. When you work with a renewable energy company, you are getting your power from renewable resources; solar energy is one of them. This sort of energy doesn't create greenhouses gas, and it doesn't contaminate water or air. It is independent and a decent way to give energy to your home or business.

✓ POWER PRICE INDEPENDENCE

When you work with a customer service company, your costs are likely on the ascent consistently. Be that as it may, when you are on the framework with a renewable energy company, you have a lot steadier rates. They use renewable energy, for example, solar energy, and that helps level the standards. Also, they regularly charge you for what you use and not variable rates that are as a result with other power suppliers.

✓ SOLAR ENERGY CREATES JOBS

When you bolster renewable energy organizations, you are making occupations that lead to the portion of frameworks that create more energy from renewable resources. It's a positive cycle that you can genuinely remain behind. The more individuals who use renewable resources, the more individuals the organizations should support the spotless energy frameworks.

✓ RESIDENTIAL OPTIONS

With regards to renewable energy resources, you can be sure that your house is being powered through household energy generation. Solar energy and other renewable resources that make your power when you're with a renewable supplier all originate from directly here in the USA. You don't need to take energy from another piece of the world to get what you require. That likewise helps hold the costs down and at steadier dimensions.

✓ ENHANCED GRID SECURITY

The more individuals who use renewable resources through a renewable energy company, the fewer power outages you will see. At the point when more individuals use the power created through common resources, the matrix will be increasingly secure. It will be more reluctant to have common or human-caused issues because it is an increasingly regular process that is more diligently to interfere.

✓ MORE CAUSES FOR UNDERUTILIZED LAND

Utilizing renewable resources like solar power will cause there to be more motivated to use arrive that has been underutilized as of not long ago. Most regions still have a ton of land if you turn away from the large urban communities and that arrive is being used in vain. With renewable resources in play, that land can make extraordinary esteem.

SOLAR ENERGY ADVANTAGES

There are numerous favorable circumstances to utilizing solar energy through a renewable energy company in both of all shapes and sizes ways. Shoppers have questions, yet as they do inquire about, they see the advantages. Numerous property holders need to filter out their inquiries first and afterward settle on a choice regardless of whether solar energy with a renewable energy company is directly for their home and family.

THE FUTURE OF SOLAR ENERGY

At first, become flushed, solar energy is maybe the most elegant solution to our energy needs. The sun shoots our planet's surface with all that anyone could need the energy to prop us up until the end of time.

The United States government assesses that the Earth gets more than 173,000 terawatts of energy consistently, which is more than multiple times what humankind needs.

The test has always been gathering that energy. Even though the vast majority know about photovoltaic cells, solar panels have been sufficiently costly to keep them solidly in the extravagance section. For quite a long time the low efficiency of solar panels and the large expenses per square inch of these panels made solar power monetarily unviable.

That has now changed. In the five years somewhere in the range of 2008 and 2013, the cost of solar panels fell by more than 50 percent. Somewhere in the range of 2015 and 2017, specialists gauge the value will fall another 40 percent. Analysts in the United Kingdom state they are amazed by how brisk solar selection is developing. They measure that the costs will fall quick enough to enable solar to contribute 20% of our energy utilization by 2027. That benchmark would have been impossible a couple of years back.

It appears the innovation has gotten up to speed as far as costs and efficiency. It's presently at the very edge of the mass selection. Be that as it may, what would be an ideal next step? What's in store for the eventual fate of solar energy?

New Business

Each innovation brings new open doors for business. Tesla and Panasonic are as of now arranging a humongous solar panel producing processing plant in Buffalo, New York. Tesla's Power divider is as of now a standout amongst the most mainstream local energy stockpiling gadgets on the planet. The vast players aren't the main ones profiting from the solar energy blast.

There is probably going to a be a great deal of interest for land. Landowners and ranchers can rent out their property for the development of new solar farms.

Enthusiasm for medium voltage link could ascend since solar farms should be newly associated with the lattice. All the new open doors will drive costs lower and stimulate the tech further.

Bio-solar cells

Scientists have explored different avenues regarding organic material in solar cells for some time now. Microbes (explicitly cyanobacteria) can, in the end, make it less demanding to power remote gadgets. The efficiency of these bio-solar cells is no place near conventional PV cells, yet there is trust the innovation will continuously make up for lost time. One of the scientists at the Binghamton University's Thomas J. Watson School of Engineering and Applied Science accepts bio-cells would be useful for remote zones where supplanting batteries much of the time isn't a choice.

Better Conversion to Electricity

Scientists from Israel and Germany banded together up to examine if there was a superior way to convert sunlight into power. It turns out that the most productive approach is additionally the most widely recognized – photosynthesis. The examination affirmed that utilizing biomass as fuel could, in the long run, enable us to make fake photosynthesis machines. These could convert sunlight into energy and store in a progressively characteristic way for later use.

Sliding Panels

A few nations come up short on the space for solar farms. An elegant solution to this issue is skimming solar farms. Ceil and Terre International, a French energy company, has been taking a shot at a vast scale, coasting, and solar solution since 2011. They have just introduced a preliminary ranch off the bank of the UK and are currently taking a gander at endeavoring comparative undertakings in India, France, and Japan.

Remote Power from Space

The Japanese Space Agency (JAXA) thinks drawing nearer to the sun is the ideal way to drive efficiency and gather more power. The group's Space Solar Power Systems (SSPS) venture is endeavoring to send solar panels to close Earth circle. The energy gathered will be remotely transmitted back to the base station using microwaves. If practical, this innovation could be a substantial distinct advantage.

Energy Harvesting Trees

A group of analysts in Finland is attempting to make a tree that stores solar energy in its leaves. These leaves could then be used to little power apparatuses and cell phones. The trees are probably going to be 3D printed, utilizing biomaterials that emulate natural wood. Each leaf produces power from sunlight, however, can likewise use dynamic energy from the breeze. The trees are intended to endure inside just as outside. The undertaking is as of now in the model stage at the VTT inquire about focus in Finland.

Better Efficiency

Efficiency is, right now, the greatest obstacle to better solar power. Right now, over 80% of every single solar panel has an energy efficiency of under 15 percent. The more significant part of these solar panels is stationary, which implies they pass up direct sunlight. A more substantial portion of the sunlight that hits the boards is squandered. Better structure, better science and the use of sunlight-retaining nanoparticles could drive efficiency.

A few specialists trust they have figured out how to catch the infrared range of light for use in solar panels. At present, infrared beams go directly through the panels and are squandered. In any case, if this range of undetectable light can be caught, it could help energy efficiency by 30 percent.

In the interim, IBM is endeavoring to make individual PV cells littler with the goal that a more considerable amount of them could be crushed into more tightly spaced. The company trusts it could, in the end, pack multiple times more PV cells into a similar space.

Solar energy is unmistakably what's to come. Till date, humankind has just touched the most superficial layer of the sun's actual potential. The sun conveys more energy to the planet's surface than what is used each year. While the costs have diminished definitely throughout the years, the innovation has continued as before. Analysts over the globe are working eagerly to enhance the way sunrays are gathered and converted into energy.

The determined drive of innovation will, in the long run, help solar energy contribute a noteworthy part in the yearly energy needs. Better and progressively productive gadgets will be powered by the sun and can store this energy for more extended periods. The next energy blast is set to change lives until the end of time.

CHAPTER THREE

THE MOST EFFECTIVE METHOD TO UNDERSTAND ELECTRICITY: WATTS, AMPS, VOLTS, AND OHMS

The four most essential physical amounts in electricity are:

- Voltage (V)
- Current (I)
- Resistance (R)
- Power (P)

Every one of these amounts is estimated utilizing distinctive units:

Voltage is rated in volts (V)

Current is estimated in amps (A)

Resistance is estimated in ohms (Ω)

Power is estimated in watts (W)

Electrical power, or the wattage of an electrical system, is continuously equivalent to the voltage duplicated by the current.

A system of water funnels is regularly utilized as a similarity to enable individuals to see how these units of electricity cooperate. In this relationship, voltage is proportional to water weight, current is proportionate to stream rate, and resistance is identical to pipe measure.

In electrical engineering, there is an essential condition that clarifies how voltage, current, and resistance relate. This condition, composed underneath, is known as Ohm's law.

Ohm's Law

V = I x R

Ohm's law expresses that voltage is equivalent to the current streaming in a circuit time the resistance of the circuit.

One method for understanding Ohm's law is to apply it to the fanciful pipes system we've utilized as a portrayal of an electrical system.

Suppose we have a tank of water connected to a hose. If we increment the weight in the tank, more water will leave the hose. Consequently, if we increment the voltage in an electrical system, we will likewise build the current.

On the off chance that we make the distance across of the hose littler, resistance will expand, making less water leave the hose. Along these lines, on the off chance that we increment the opposition in an electrical system, we will diminish the current.

With this concise presentation of the operations of an electrical system, how about we bounce into every one of the units of electricity independently and find out about them in more detail.

The picture above delineates a primary electrical circuit with a bulb, some wire, and a battery.

WHAT ARE VOLTS?

Volts are the base unit used to gauge Voltage. One volt is characterized as the "distinction in electric potential between two points of a directing wire when an electric current of one ampere disseminates one watt of power between those focuses." The volt is named after the Italian physicist Alessandro Volta.

In our battery chart over, the battery gives what is known as a potential contrast in an electric circuit, or voltage. If we return to our water similarity, the cell resembles a water siphon that pushes water through a pipe. The siphon expands the weight in the pipeline, making the water stream.

In electrical engineering, we call this electrical weight voltage and measure it in volts. A voltage of three volts can be composed as 3V.

As the quantity of volts expands, the current increments as well. In any case, all together for the present to stream, the electrical conduit or wire must circle back to the battery. On the off chance that we break the circuit, with a switch, for instance, no current will stream.

There are standard voltage outputs for regular items like batteries and household outlets. In the United States, the standard voltage output for a household outlet is 120V. In Europe, the standard voltage output for a household outlet is 230V. Other standard voltage outputs are recorded in the table beneath.

Regular Voltages

Item

Voltage

Single-cell, rechargeable battery

1.2V

Single-cell, non-rechargeable battery

1.5V– 1.56V

USB

5V

Car battery

2.1V per cell

Electric vehicle battery

400V

Household outlet (Japan)

100V

Household outlet (North America)

120V

Household outlet (Europe, Asia, Africa, Australia)

230V

Quick travel third rail

600V– 750V

High-voltage electric power lines

110,000V

Lightning

100,000,000V

WHAT ARE AMPS?

The ampere, frequently abbreviated to "amp" or An, is the base unit of electric current in the International System of Units. It is named after the French mathematician and physicist André-Marie Ampère, who is viewed as the dad of electrodynamics.

Electricity comprises of the stream of electrons through a conduit, for instance, an electric wire or link. We measure the rate of a flow of power as an electric current (similarly as we think about the speed of the stream of water in a waterway as the waterway current). The letter used to speak to flow in a condition is I.

Electric current is estimated in Amperes, abbreviated to Amps or the letter A.

A current of 2 Amps can be composed as 2A. The higher the current, the greater electricity is streaming.

The International System of Units (SI) characterizes amps as pursues:

"The ampere is that consistent current which, whenever kept up in two straight parallel conductors of interminable length, of insignificant roundabout cross-segment, and set one meter separated in a vacuum, would deliver between these conductors a power equivalent to 2×10^{-7} newton per meter of length."

Electric Current Demonstration

WHAT ARE OHMS?

Ohms are the base unit of resistance in an electrical system. The ohm is characterized as "an electrical resistance between two of a conduit when a consistent potential contrast of one volt, connected to these focuses, delivers in the channel current of one ampere, the transmitter not being the seat of any electromotive power." The ohm is named after the German physicist Georg Simon Ohm.

Resistance is estimated in ohms, or Ω (omega), for short. Along these lines, five ohms can be composed 5ω.

In our battery outline above, on the off chance that we expelled the bulb and reconnected the wire to the battery was short-circuited, the wire and battery would get exceptionally hot, and the battery would before long be level because there would be no resistance in the circuit. With no opposition, an immense electrical current would stream until the point when the battery was unfilled.

When we add a bulb to the circuit, resistance is made. There is presently a neighborhood "blockage" (or narrowing of the pipe, per our water pipe relationship) where the current encounters some resistance. This incredibly diminishes the current streaming in the circuit, so the vitality in the battery is discharged all the more gradually.

As the battery powers the current through the bulb, the battery's vitality is discharged in the bulb as light and warmth. The current conveys put away energy from the array to the bulb, where it is transformed into light and warmth vitality.

The picture above demonstrates a light as the primary driver of electrical resistance.

WHAT ARE WATTS?

A watt is the base unit of power in electrical systems. It can likewise be utilized in mechanical systems. It gauges how much vitality is discharged every second in a network. In our battery graph, the extent of both the voltage and the current in the bulb decide how much spirit is released.

In the graph over, the light would get more splendid as the power, estimated in watts, increments.

We can ascertain the power discharged in the bulb and of the electrical system all in all, by increasing the voltage by the current. In this way, to compute watts, the accompanying recipe is utilized.

HOW TO CALCULATE WATTS

W = V * I

For instance, a current of 2A moving through a bulb with a voltage of 12V crosswise over it creates 24W of power.

How to Calculate with Watts, Amps, Volts, and Ohms

On the off chance that you need to complete an electrical figuring including voltage, current, resistance, or power, reference the formulae hover underneath. For instance, we can ascertain the power in watts by referencing the yellow zone in the circle.

This formulae circle is precious for some, electrical engineering assignments. Keep it convenient whenever you are managing an electrical system.

The following are some precedent conditions that are unraveled utilizing the formulae.

Precedent Equations

1. What is the current in an electrical circuit with a voltage of 120V and 12ω of resistance?

$I = V/R$

$I = 120/12$

$I = 10A$

The current in an electrical circuit with a voltage of 120V and 12ω of resistance is 10A.

2. What is the voltage over an electrical circuit with a current of 10A and 200ω of resistance?

$V = I \times R$

$V = 10 \times 200$

$V = 2000V$

The current in an electrical system with 10A and 200ω of resistance is 2000V.

3. What is the resistance in an electrical system with a voltage of 230V and a current of 5A?

R = V/I

R = 230/5

R = 46ω

The resistance in an electrical system with 230V and 50A is 46ω.

The most effective method to USE A MULTIMETER TO MEASURE VOLTAGE, CURRENT, AND RESISTANCE

WHAT IS A MILLIMETER?

A computerized millimeter or DMM is a valuable instrument for estimating voltage, current, and obstruction, and a few meters have an office for testing transistors and capacitors. You can likewise utilize it for checking coherence of wires and wires. On the off chance that you like to DIY, do vehicle support or investigate electronic or electrical hardware, a millimeter is a helpful assistant to have in your home toolbox.

Volts, Amps, Ohms - What Does everything Mean?

Volts

This is the weight in an electrical circuit.

Amps

This is a proportion of the flow streaming in an electrical circuit.

Ohms

A proportion of the protection from a stream in a circuit

Voltage Source

This delivers a present stream in a circuit. It could be a battery, convenient generator, mains supply to a home, alternator on your vehicle motor or seat power supply in a lab or workshop.

Load

A gadget or part which draws power from a voltage source. This could be an electronic resistor, knob, electric radiator, engine or an electrical machine.

Ground

This usually is the point in a circuit to which the negative terminal of a battery or power supply is associated.

DC

Coordinate current. Current streams just a single route from a DC source, a case of which is a battery

Air conditioning

You are substituting Current. Current streams one path from a source turns around and afterward streams the other way. This happens all the time at a rate dictated by the recurrence which is usually 50 or 60 Hertz. The mains supply in a home is AC

Extremity

A term used to portray the bearing of the stream of current in a circuit or which focuses are sure and which are contrary writ a reference point

Utilizing a Multimeter - Measuring Functions on the Instrument

An essential multimeter encourages the estimation of the accompanying amounts:

DC voltage

DC

Air conditioning voltage

Obstruction

Congruity - demonstrated by a bell or tone

Some fundamental meters don't have an AC range.

Moreover, meters may have the accompanying capacities:

Capacitance estimation

Transistor HFE or DC gain

Temperature with an extra test

Diode test

Recurrence

The esteem estimated by the instrument is demonstrated on an LCD show or scale. Research center seat DMMs sometimes has seven fragments LED shows.

Solar installations are getting less demanding constantly, and there's a lot of do-it-without anyone else's help data out there. Be that as it may, would you say you are prepared to go the DIY course?

In case you're keen on solar power, without a doubt you realize that solar electricity is useful for the earth, national security, and the air we inhale, also your electricity bill. Also, that it's a standout amongst the ideal approaches to diminish your family unit's commitment to a dangerous atmospheric division. You've likewise most likely heard that going solar can be less expensive than paying for utility power, and you may ponder whether this case is valid. Indeed, much of the time, it is accurate. It just requires investment for the gradual reserve funds to surpass the underlying venture (from that point forward, the solar power is free). On the off chance that you introduce the solar system yourself, you can hit this tipping point a great deal sooner — at times, in a fraction of the time.

That conveys us to the following central issue: Can you indeed introduce your solar panels? Once more, the appropriate response is yes. If you can drive slack jolts and gather pre-assembled parts, and in case you're willing to go through multi-day or two on your rooftop (or not, in case you're mounting your panels on the ground), you can introduce your solar system. You don't need to realize how to connect the solar panels to your family unit electricity or the utility framework. You'll employ a circuit maintenance specialist for the house hookup, and the utility organization will deal with the rest, for the most part for nothing. For an off-lattice system, the utility organization isn't associated with anyway.

Maybe disappointingly, this activity isn't even a decent reason to purchase new power apparatuses, since the just a single you require is a suitable bore.

Things being what they are, if this is such a feasible undertaking, for what reason do the vast majority utilize expert installers? First off, many individuals have valid justifications for employing out everything, from oil changes to shopping for food. Solar experts handle more than the installation. They plan the system, they apply for discounts and credits, they arrange all the essential parts, and they get the grants and pass every one of the examinations. The truth of the matter is, you can do these things yourself if you have an accommodating counsel and you are eager to pursue the standards of the local building authority (that is the place you'll get those grants).

Solar installations are getting less demanding constantly, and you may be shocked at what amount do-it-without anyone else's help (DIY) help is accessible. Two good precedents are Watts and the Database of State Incentives for Renewables and Efficiency (DSIRE). PVWatts is an online mini-computer that encourages you to measure a solar-electric system dependent on the area and position of your home and the point of your rooftop. Solar masters utilize a similar straightforward instrument, yet it's free for everybody. DSIRE offers and a la mode, far-reaching posting of sustainable power source discounts, tax cuts, and other budgetary motivating forces accessible in any zone of the United States. Furthermore, it's additionally free and straightforward to utilize.

Those two resources alone help answer the two most basic inquiries homeowners have about solar electricity: How enormous of a system do I require? What's more, How much will it cost? Different resources incorporate solar gear providers that take into account DIYers and offer to obtain and specialized help, just as shopper inviting industry sources like Home Power magazine and the online network Build It Solar. What's more, there's no law that says DIYers can't contract a solar expert for help with specific parts of their undertaking, for example, making structure particulars, picking hardware, or getting ready allow archives.

We ought to likewise say in advance that introducing your solar panels is certifiably not a procedure all around served by compromising. We don't need you to organize your system without an allow or without enlisting a circuit maintenance specialist to make the last hookups. (Indeed, even expert solar installers use circuit testers for this stuff.) The allow procedure can be agony, honestly, however, it's there to guarantee that your system is sheltered, for you as well as for crisis responders who may need to work around your smaller than a typical power plant. When you work with the local building division, you additionally find out about fundamental plan factors, for example, wind and snow stacks, thatare explicit to your territory

It's time for the litmus test that reveals to you whether to continue intensely as an amateur solar installer or to give over control to an expert. For the vast majority of you, the choice will come down to the tenets of the local building authority (probably your city, province, township, or state) or your utility provider, both of which may necessitate that solar installations be finished by an authorized proficient. This is additionally the best time to affirm that your undertaking won't be nixed by your zoning office, chronicled region benchmarks, or your homeowner's association.

Amateur installation is allowed by the local building authority and your utility provider.

Prerequisites for amateur installation are sensible and satisfactory. A few experts require nonprofessionals to breeze through tests exhibiting essential information of electrical and other family unit systems, yet such criteria may not be broad.

You're alright with a few hours of physical housetop work (those with ground-mount systems get a go here), AND you're savvy enough to wear authentic fall-capturing hardware (not a rope tied around your midsection). You may feel as confident as Mary Poppins moving on housetops, yet she can fly; you ought to be fastened.

You don't live in a recorded area or, if you do, the zoning authority grants PV systems (with adequate limitations).

Your homeowner's association, on the off chance that you have one, grants PV systems (with worthy limitations). Sometimes the homeowner's association may require a touch of poking to give consent.

You have a standard sort of material (black-top shingles, standing-crease metal, wood shingles, standard level rooftop). If you have slate, solid tile, dirt tile, or other delicate/claim to fame material, counsel a material expert as well as contract out the PV installation, this isn't a significant issue.

Cautioning: PV systems are innately dangerous and conceivably destructive. As a DIY system installer and proprietor, you should comprehend, regard, and moderate the dangers required with all installation and support assignments. Give careful consideration to security admonitions just as all necessities in the local building and electrical codes and gear guidance manuals.

For homeowners thinking about introducing solar panels, a vital thought is regardless of whether the home's solar panel system ought to be associated with the lattice (network tied) or of the matrix.

Albeit off-lattice homes were progressively regular previously, network tied homes are expanding as the reasonableness and fame of solar systems increment.

CHAPTER FOUR

WHAT IS A GRID-TIED SYSTEM?

A grid-tied solar panel system is a solar energy system that is associated with the electrical grid and, in this way, utilizes power from both the solar panel system and the electrical grid. Along these lines, a grid-tied solar system doesn't need to meet the majority of the power requests of the home.

If necessary, the home can draw energy from the grid now and again, (for example, on overcast days or around evening time) when solar panels are not delivering at full effectiveness. Moreover, if more energy than is required is produced by the solar panels of one home, that abundance energy will be bolstered into the grid for use somewhere else.

INTERFACING YOUR HOME TO THE GRID

Interfacing your home to the grid will require an association from you and the provider of your solar panel system.

First of all, your solar system provider must know the neighborhood interconnection laws. The interconnection laws are standards and methods that explicitly apply to circumstances where a sustainable power source system, for example, a solar energy system is "stopped" into the power grid. The interconnection laws express the terms that must be trailed by both solar energy system proprietors and utilities.

To begin with a grid-tied system, your solar system provider will record interconnection and net metering applications to the service organization.

ADVANTAGES OF A GRID-TIED SYSTEM

A grid-tied solar system has a few noteworthy focal points over off-the-grid solar systems:

Unwavering quality: Solar panel systems are not impeccable. There will undoubtedly be days when productivity isn't what it could be, and the system doesn't deliver enough. Be that as it may, a large number of your day by day (and daily) exercises will keep on requiring the utilization of power.

Not at all like off-grid are systems which may come up short on power, grid-tied solar systems less inclined to abandon you out of the loop at previous occasions. If the power created from your solar panel system is not exactly ideal, additional energy will be pulled from the grid. The grid goes about as reinforcement for your solar energy system.

Less energy is squandered subsequently, and the productivity of your solar power system goes up.

Except in case of a power blackout, you will dependably approach power amid whenever of day, as long as your order is associated with the grid.

Expenses: To work legitimately, off-grid solar energy systems require increasingly real hardware that gets costly rapidly. Less gear, for the most part, implies bring down establishment and upkeep costs. This happens to be the situation with most grid-tied systems. Since the power grid works as a battery for your system, you don't need to pay for batteries; you don't need to pay for the upkeep that is included with those batteries.

Net metering: The key idea to comprehend about a grid-tied system is that it enables you to nourish power to the grid amid the day, when you might deliver abundance energy, and to utilize the grid supply during the evening. Net metering is a charging procedure that credits the proprietors of grid-tied systems when they create more energy than the home needs.

Since grid-tied homes are typically net-metered, the power meter tracks this trade between your solar system and the grid. Overabundance energy age prompts your power meter turning in reverse instead of forward, along these lines giving you credit. The credit can be utilized to counterbalance installments for future power use.

OFF GRID SOLAR SYSTEM

An off-grid solar system isn't associated with the electricity grid and hence requires battery stockpiling. An off-grid solar system must be appropriately structured with the goal that it will create enough power consistently and have enough battery ability to meet the home's prerequisites, even in the profundities of winter when there is less daylight.

The tremendous expense of batteries and inverters implies off-grid systems are substantially more costly than on-grid systems as are typically just required in progressively remote zones that are a long way from the electricity grid. Anyway battery costs are decreasing quickly, so there is currently a developing business sector for off-grid solar battery systems even in urban communities and towns.

There are distinctive sorts of off-grid systems which we will broadly expound later, however, for the present. This depiction is for an AC coupled system; in a DC combined system control is first sent to the battery bank, at that point sent to your machines

The battery bank: In an off-grid system, there is no open electricity grid. When solar power is utilized by the apparatuses in your property, any overabundance power will be sent to your battery bank. When the battery bank is full, it will quit getting power from the solar system. At the point when your solar system isn't working (evening or overcast days), your machines will draw control from the batteries.

Backup Generator: For times of the year when the batteries are low on charge, and the climate is exceptionally shady you will, for the most part, require a backup control source, for example, a backup generator or gen-set. The span of the gen-set (estimated in kVA) ought to be satisfactory to supply your home and charge the batteries in the meantime.

Off-grid solar frameworks can run autonomously from the electric grid. To achieve this, they require additional equipment. DC power produced by the PV boards is fed into a charge controller, which directs the charge rather than the utility company's PC controlled grid. From that point, it's fed into a DC battery bank, where it's put away. The charge controller decides if to stream charge, full-stack charge, or prevent current from over-burdening the battery bank dependent on the battery charge level. Your home interfaces with the battery bank through a power inverter, which changes over the DC control from the batteries to the 120V AC control utilized by most family outlets. Even though lead-acid vehicle batteries (particularly deep cycle) can be used to fabricate a battery bank, it's not prescribed, as these batteries have a short lifespan, particularly when employed day by day. Also, these batteries lose around 20% of the putaway vitality, squandering significant power and undermining the earth neighborly part of solar. Numerous organizations create lithium-ion batteries explicitly intended for solar applications, similar to the prevalent Tesla

Powerwall, a 6.4 kWh Li-ion battery with a 10-year lifespan. These batteries are a lot nearer to the 7% vitality waste made by the utility company's grid. However, the effectiveness diminishes over the lifespan of the cell. New kinds of batteries, similar to the zinc bromide ZCell, may enhance productivity after some time. However, they still can't seem to be tried in the private market. Remember that some vitality is lost in charging the battery, paying little mind to what sort of cell is utilized. As a result of this additional gear, off-grid solar is more costly than grid-tied with the Tesla Powerwall costing $3,000 (or $6,000 amid the 20-year lifespan of the solar boards) and a 10kW inverter costing $300-$500. An additional DC separate switch is likewise vital between the battery and inverter, which includes an extra $100-$200. Off-grid solar is perfect for remote places or immature territories where the electrical grid isn't steady. It's additionally an incredible solution for individuals who can manage the cost of the direct expenses, what's more, wish to untether themselves from the electric grid. A drawback to off-grid solar winds up visible when daylight isn't as promptly accessible, for example, amid great storms in the winter, when the sun is as of now up for a shorter range of time every day.

THE ADVANTAGES OF GRID-TIED, OFF-GRID, AND HYBRID SOLAR POWER SYSTEMS

There are three noteworthy sorts of solar power systems that can be utilized in Frederick and these are grid-tied, off-grid, and a half breed. Everyone has its very own arrangement of favorable circumstances. Peruse on to discover so you can pick the ideal one for your home.

Grid-Tied

The grid-tied system keeps your home associated with the grid regardless of having a green wellspring of energy. Its points of interest include:

- **Increasingly practical**

The grid-tied system will enable you to have more funds since it is useful, it offers net metering, and it requires a minimal effort of gear and establishment. It doesn't need the utilization of batteries, which are typically costly. Also, since there are no batteries, upkeep cost is kept at the very least. Generally speaking, the grid-tied system is shabby to introduce and keep.

- **The grid fills in as your virtual battery.**

Power is an asset that ought to be expended progressively. Nonetheless, this isn't generally the situation, so there is regularly a need to store it somewhere else before utilizing it. With the grid-tied system, the whole grid fills in as your virtual battery where abundance power can be kept incidentally. Since the grid fills in as a battery and there indeed is no wasteful battery included nothing goes to squander in this system.

Off-Grid

At the point when your home does not approach the grid, your next best alternative is to pick an off-grid system. It nearly has the same system from the grid-tied however as opposed to having a grid to interface with, it stores power in a battery. Here are the benefits of this system:

- **A superior alternative for homes with no entrance to the grid**

Rather than having miles and miles of power lines installed in your home to approach the grid, the off-grid alternative is better. It is less expensive than having the power lines installed while as yet giving nearly same dependability from a grid-tied system.

- **Your home progresses toward becoming energy adequate**

Some time ago, if your home did not approach the grid, there was no alternative for it to be energy adequate. With the off-grid system, you can have power day in and day out, gratitude to the battery that stores it for you. Having sufficient energy for your home includes a layer of security. Also, you will never be influenced by power disappointments since you have an independent source at home.

Half and half

The half and half system consolidate the best out of the grid-tied and off-grid systems.

- **More efficient than the off-grid system**

The half and half system are more affordable to introduce and keep up contrasted with the off-grid system. With it, you don't have to put a reinforcement generator. Besides, you can diminish the measure of your battery. Also, the off-top power from the grid is a lot less expensive contrasted with diesel.

- **It is a promising system.**

Even though not a ton of homes has a half breed system, it has a great deal of potential. It is a decent progress system to shrewd grids of things to come.

CHAPTER FIVE

PHOTOVOLTAIC SOLAR

Solar panels are more perplexing than meets the eye. Photovoltaic (PV) cells are the components that convert daylight into electricity. These cells are organized into PV modules, all the more ordinarily alluded to as solar panels, which electrically interface all the cells to increase the all-out power yield. Modules can be associated together to frame a PV cluster.

TYPES OF SOLAR PV SYSTEMS

PV-Direct Systems: These systems are the most straightforward because they neither include a battery nor are they associated with the utility grid. They can create electricity when solar energy is accessible.

Off-Grid Systems: These systems are not associated with the utility grid and create all of the building's electricity. Abundance energy generated by these systems is stored in batteries and after that in this manner utilized around evening time or in cold climate when solar energy isn't accessible. Off-grid systems are generally utilized in territories without utility administration. They require a battery bank, a charge controller, an inverter, and detaches.

Grid-Tied Systems with Battery Backup: These systems work equivalent to an off-grid system aside from that they are associated with the grid. If the system can't give enough energy through the PV system or the battery, the utility will supply power.

Battery-less Grid-Tied Systems (Grid-Direct, Grid-Interactive): Battery less grid-tied systems are the most widely recognized PV application. These systems are associated with the utility power grid and utilize the utility grid as a reinforcement wellspring of power. If more energy is required than is being created by the solar PV system, the service organization supplies the distinction. On the off chance that more energy is being produced than needed, the abundance streams in reverse through the electrical meter to the service organization, where others then utilize it. These systems required inverters and required electrical wellbeing gear.

TYPES OF PV PANELS

Single-gem (monocrystalline) modules: These cells are the most effective and most costly PV cells since they will result in general believer more daylight energy to electricity than different types of PV cells. They are made of silicon and are cut from cylindrical ingots that take after a huge, round salami.

Multicrystalline (polycrystalline) modules: These cells are marginally less effective that single-precious stone cells. They are made of silicon cut from different precious stones.

Thin-film modules: Thin-film PV modules have the most reduced effectiveness but at the same time are bring down in expense per watt of energy produced. They are fabricated using splash on or print-on strategies, instead of forming cells from ingots or squares of liquid silicon. They are manufactured using numerous types of metals including silicon, gallium, cadmium, tellurium, and copper.

Extra Information

American Solar Energy Society – Covers an assortment of points including how solar electric systems function, what PV cells are made of, and unique adaptations of commercially available PV panels.

Solar Electricity Basics and Solar Electric System Types – Provides depictions of various solar electric systems, including batteryless grid-tied systems, the most well-known type of private arrangement.

Small Solar Electric Systems – Provides a review of the technical subtleties of solar PV cells and modules, semiconductor materials, home solar system components, and thin film solar cells.

HYBRID SOLAR SYSTEM

Hybrid solar systems produce power in the same path from a typical grid-tie solar system yet use batteries to store energy for later use. This capacity to save energy empowers most hybrid systems to likewise work as a reinforcement power supply during a power outage, like a UPS system.

Traditionally the term hybrid alluded to two age sources, for example, wind and solar yet more as of late the term 'hybrid solar' alludes to a combination of solar and battery stockpiling which not at all like off-grid systems is associated with the electricity grid.

Essential design graph of a typical solar hybrid system (DC coupled battery)

Why store solar energy in a battery?

Numerous administrations and system administrators have diminished the solar feed-in levy or FiT (money or credit got for feeding solar energy to the grid). This implies conventional grid-feed solar systems have turned out to be less alluring as the vast majority are working during the day and not home to utilize the solar energy as it is created, along these lines the energy is nourished into the grid for almost no arrival.

A solar hybrid system stores your abundance solar energy and can likewise give back-up power during a power outage. This is ideal for property holders although, for most of the businesses which work during the light hours, a typical grid-feed solar system is as yet the most prudent decision.

Hybrid solar systems empower you to store solar energy and use it when you're home during the evening when the expense of electricity is typically at the pinnacle rate.

The capacity to store and utilize your solar energy when wanted is alluded to as self-use or self-utilization. It works in the same route from an off-grid power system however the battery limit required is far less, usually sufficiently only to cover top utilization (8 hours or less) rather than 3-5 days with a standard off-grid system.

SOLAR HYBRID ADVANTAGES

Allows utilization of solar energy during pinnacle times (self-use or load-shifting)

Power accessible during a grid blackout or power outage – UPS work

Empowers propelled energy the executives (i.e., top shaving)

Empowers energy independence

Diminishes power utilization from the grid (decreased interest)

HYBRID SOLAR SYSTEM TYPES

This is a technical guide to the different hybrid solar systems and inverters available. The systems accessible may change depending on the nearby solar merchants and electricity arrange administrators in your nation. Allude to our total hybrid and battery stockpiling audit for an immediate examination of the different hybrid inverters and battery stockpiling systems.

Hybrid systems can be ordered into two main types:

1. All-In-One Hybrid Inverter/System

The most prudent hybrid solar system utilizes an all-in-one hybrid inverter which contains a solar inverter and battery inverter/charger together with astute controls which determine the most productive utilization of your available energy.

An all-in-one hybrid system is a hybrid inverter together with a lithium battery in one complete bundle, usually about the extent of an ice chest. Anyway like most machines there are numerous highlights and capacities which separate the vast assortment of hybrid systems accessible.

- **1A. All-in-one inverter (no back-up)**

This is the most fundamental kind of hybrid solar inverter and works much like a grid feed solar inverter yet, also, empowers stockpiling of solar energy in the battery system for self-use. The main detriment of this type of Inverter is that does not contain a grid separation gadget which implies it can't supply power when there is a power outage (customarily known as an uninterruptible power supply or UPS work). Although if grid dependability isn't an issue, this straightforward hybrid inverter would be a decently affordable decision.

ACCESSIBLE INVERTERS: Solar X-hybrid, Growth, Solar Edge, Sun develop, (Samsung hybrid system including batteries)

All-in-one hybrid inverters...

- **1B. All-in-one inverter with back-up (UPS)**

This further developed all-in-one hybrid inverter can regularly be utilized as both an on-grid or off-grid inverter as it has back-up capacity worked in. Under ordinary task, it can supply power to the home (assigned power circuits), charge the batteries and abundance power can be nourished into the grid. If there is a power outage or the grid winds up unsteady the unit will automatically change over to the battery supply and continue to work independently from the electricity grid (in usually not exactly a large portion of a second).

ACCESSIBLE INVERTERS: Red back Technologies, Solar X-hybrid E arrangement, Omnik, Good we ES arrangement, SonnenBatterie Eco, Solar Edge

- **1C. All-in-one system with integrated battery**

A later pattern is to bundle the all-in-one hybrid inverter together with a battery system in one complete unit. This offers an extraordinarily perfect and cost impact choice which is usually about the span of a medium ice chest. These systems are exceptionally accurate and straightforward to install however can have a few impediments as a few models can't be extended at a later date.

Accessible SYSTEMS: Red back Technologies inc battery, Solar X-Box, Alpha ESS, Samsung ESS, Fronius bundle, Power wall 2 (mid-2017)

Crossover systems with incorporated battery

2. Framework Interactive Hybrid And Off-Grid Systems

As of not long ago (before less expensive across the board half breed inverters) most cross breed systems comprised of two unique inverters which cooperated to frame what is known as an AC coupled system; a standard solar inverter and a complex intuitive or multi-mode battery inverter.

The Solar inverter can be any standard unit yet it is generally either a similar brand or is good with the intelligent inverter to advance battery charging.

The intelligent or multi-mode inverter goes about as a battery inverter/charger and full energy the executive's system, utilizing shrewd programmable programming to advance energy use. Smart inverters supply power similarly as an off-network inverter yet also control framework association (import and fare power) and can be set up to begin and run a back-up gen-set (generator) naturally.

Key highlights of an intelligent lattice system

Powerful battery inverter to supply ceaseless high loads

Progressed multi-arrange charger for lead-corrosive or lithium batteries

Programmed AC exchange switch (UPS work) worked in

High go through power capacity.

High flood stack capacity

Generator control – Auto begins/stop.

Can likewise be utilized for the top of the line Off-lattice power systems

Remote checking

Network feed-in and constraining when abundance power is created

Peak shaving to diminish peak request

The load was moving and propelled energy request the board.

Progressed intuitive systems are utilized for off-lattice and crossover establishments which require an abnormal state of power the board. The dominant programming used to run Interactive inverters empower energy controls, for example, peak shaving, in addition to information logging and PLC abilities through exceptional info/yields and hand-off controls. These systems can likewise work with substantial battery banks and fuse particular battery checking and temperature sensors to drag out battery life.

Because of the numerous highlights and propelled programming the cost of intuitive inverters usually is higher than across the board inverters however in multiple applications the additional cost merits the other venture as they are commonly progressively robust, increasingly active and empower future extension.

CHAPTER SIX

FACTS ABOUT SOLAR ENERGY

The U.S. solar industry has seen noteworthy development in the previous decade while the cost of solar has declined by almost 70 percent. Since solar has entered the standard, regular homeowners are beginning to think about how much solar could spare them and how straightforward doing the switch indeed could be. In case you're starting to consider installing solar panels, it's useful to comprehend the 10,000-foot view of solar power.

✓ **Solar is presently the least expensive and most rich energy source on the planet**

In December 2016, the cost of building and installing new solar power age dropped to $1.65 per watt, barely prevailing over its inexhaustible partner wind ($1.66/Watt) and its petroleum product rivals.

A unique defining moment regarding the financial aspects of solar versus petroleum products happened in 2016 when a solar business supplier in Dubai offered solar power available to be purchased at 0.029 pennies per kilowatt hour, setting a world record for solar just as all energy sources. Today, there are 89 Pet watts (PW) of potential solar energy generation accessible on earth, making solar the world's most rich available source of power.

✓ Over a million solar systems have been installed in the U.S. alone

In mid-2016, the millionth solar system was installed in the U.S., indenting an achievement that took 40 years for the photovoltaic business to reach. In any case, the more prominent story that accompanied this accomplishment is the anticipated course of events for the following million establishments, which is relied upon to occur in the following two years. This examination is a significant outline of solar's quick pace as the fastest developing energy resource on the planet.

✓ Various producers offer a solar panel today over 20 percent productivity

Solar panel productivity levels have been expanding as fast as solar costs are declining, on account of established researchers' emphasis on the requirement for advancement in solar innovation. To encourage offer point of view, only five years back the most productive solar panel that cash could purchase was

17.8 percent. In 2018, homeowners can get reasonable statements for solar panels in the 20 to 23 percent productivity to extend anyplace in the U.S. As far as solar cell proficiency; the two driving brands are Sun Power and Panasonic.

- ✓ **Homeowners in the U.S. have accomplished breakeven point with solar in as short as three years**

The cost of solar has dove while the loss of lattice power has proceeded to bit by bit rise, and the idea of the solar "earn back the original investment point" with solar has turned out to be increasingly alluring. In 2018, most homeowners were seeing compensation periods somewhere in the range of five and eight years, and 20-year reserve funds gauge upwards of $20,000. A few homeowners are seeing earn back the original investment indicates as low as three-four years in states where utility costs are highly similar to Massachusetts and New York.

✓ 5. The cost of a solar establishment is currently at or underneath $3 per watt in individual U.S. states

Not by any means ten years back the cost of an installed solar system was upwards of $8 a watt, and many guessed about the day when solar could break the $4/watt edge. Presently in 2018, we see the $3.00/Watt stamp produce results – cites with valuing beneath $3.00 are coming in on the Energy Sage Marketplace consistently. The standard cost per watt in 2018 is $3.16 per watt on Energy Sage, implying that a conventional estimated system (5,000 watts) will cost $11,060 after the solar ITC endowment.

✓ Planes can fly far and wide while running altogether on solar energy

In spite of the fact that many might know that solar energy can power trains, autos, and even space stations, many were doubtful when Bertrand Piccard chose to fly a solar-powered plane the world over with no extra power source than the sun. In mid-2016, the Swiss pilot and expert adventurer left from Abu Dhabi in the acclaimed airship known as Solar Impulse II, making his eagerly awaited overall return in July. The worldwide flight offered various photograph openings and created an impression around the globe about the extensive capability of solar energy.

✓ Homeowners don't need to introduce their solar panels to go solar

Individuals are regularly astonished to discover that going solar does not include installing solar panels on your property. In 2018, the idea of shared solar or community solar – installing a sizeable solar ranch from which hundreds or even a considerable number of individuals can source their power – is hugely taking off.

Community solar is currently offered by numerous vast utilities that have a motivating force to source a specific level of their provided power from inexhaustible sources. As of now, community solar is most prevalent in 4 states: California, Colorado, Minnesota, and Massachusetts. Be that as it may, with the numerous down to earth and moderate parts of community solar, the idea is rapidly picking up prevalence the nation over.

✓ **Solar energy can give power 24 hours every day.**

One of the fundamental concerns voiced by homeowners while considering going solar is, "The thing that would I do during the evening?" This is the resource's most clear boundary towards achieving official status, and solar energy stockpiling suppliers are noting the call. Various very much respected brands have entered the solar storage room (counting Tesla, LG, and Mercedes) and new challenge and development are making the cost of a solar stockpiling plunge. In 2018, homeowners can buy solar-in addition to capacity systems and be totally energy free. To become familiar with capacity, look at the best solar batteries accessible in 2017.

By and large, these eight actualities offer various points on solar's development in the ongoing decade and the manner in which it has turned into a genuine contender of petroleum product resources in 2018. In case you're thinking about a solar panel system sooner rather than later, look at a few hints for how solar customers can ensure they'll get the least cost and best hardware with their establishment:

TIPS FOR SOLAR SHOPPERS

1) Mortgage holders who get various statements spare 10% or more

Likewise with any expensive purchase, looking for a solar panel establishment takes a great deal of research and consideration, including a careful survey of the companies in your general vicinity. A recent report by the U.S. Division of Energy's National Renewable Energy Laboratory (NREL) recommended that consumers compare whatever solar number options as could be expected under the circumstances to abstain from paying expanded prices offered by the large installers in the solar business.

To locate the littler contractors that typically offer lower prices, you'll have to utilize an installer arrange like Energy Sage. You can receive free statements from verified installers local to you when you enlist your property on Solar Marketplace – mortgage holders who get at least three statements can expect to spare $5,000 to $10,000 on their solar panel establishment.

2) The greatest installers typically don't offer the best price

The higher isn't in every case better mantra is one of the fundamental reasons we emphatically encourage property holders to consider the majority of their solar options, not merely the brands large enough to pay for the most publicizing. A recent report by the U.S. government found that large installers are

$2,000 to $5,000 more costly than little solar companies. On the off chance that you have offers from a portion of the enormous installers in solar, ensure you compare those offers with statements from local installers to guarantee you don't overpay for solar.

3) Comparing all your hardware options is similarly as critical

National-scale installers don't merely offer more expensive rates – they likewise will, in general, have less solar gear options, which can significantly affect your system's electricity production. By collecting a various exhibit of solar offers, you can compare costs and reserve funds dependent on the several hardware packages accessible to you.

There are various factors to consider when searching out the best solar panels available. While specific panels will have higher efficiency evaluations than others, putting resources into first class solar gear doesn't generally result in more top reserve funds. The best way to locate the "sweet spot" for your property is to assess sites with differing gear and financing offers.

FACTORS THAT AFFECT SOLAR PV SYSTEM EFFICIENCY

Vitality efficiency factors must be carefully considered while structuring any solar PV systems if you need to get the best out of your endeavors and speculation. On the off chance that you have appliances that are not very vitality efficient you will require a reasonably large PV system (and large mark in the bank balance as well!). It doesn't bode well, regardless of whether you are incredibly wealthy. The other power source such as solar is considered because petroleum product is filthy and isn't regularly enduring (taking a gander at the running pace of increase in vitality consumption across the globe). In this manner, you might want to utilize it in an ideal way.

Be that as it may, even after you have replaced the electrical load with the most efficient appliances, despite everything you need to remember inefficiencies of the PV system which are continually hiding near. Hence, it pays to know about various factors that can conceivably corrupt your system, with the goal that you can endeavor endeavors to limit them comfortable arranging stage. Here are six vital considerations.

✓ Cable Thickness

We, for the most part, have electrical appliances working at 220V which is significantly higher compared with the typical PV system DC voltages of 12V, 24V or 48V. For a similar wattage, much higher currents are associated with the PV systems. This brings into picture resistance misfortunes in the wiring.

Give us a chance to perceive how it very well may be significant.

20 meter is the length of cable between the panel and the charge controller. A typical cable with 1.5 sq mm cross-section has a resistance of about 0.012 ohms per meter of wire length. So a 20-meter long wire will offer resistance of 20 x 0.012 = 0.24 ohms.

If it is a 24V system and a 10-ampere current is moving through this wire, at that point from the Ohm's law (V = IxR), we can calculate the voltage drop across this wire: 2.4V. It implies the voltage at the charge controller end of the cables will be 2.4V not exactly the voltage produced by the panels if a 10 Amp current is streaming. This 10% voltage drop is unacceptable.

Imagine a scenario where we utilize a six sq mm cross section cable which has a resistance of 0.003 ohms per meter. The total strength for 20-meter long cable will currently be 0.06 ohms; and the voltage drop, 10×0.06 or 0.6V. It is 2.5% voltage drop for a 24V system which may be acceptable. Be that as it may, shouldn't something be said about the increased cost of thicker cable? Similarly, there would wire all around, and careful attention must be paid to know the impact on in general system efficiency. In this way, cable length and size needs careful attention comfortable arranging stage.

Another approach to reducing resistance misfortune is to raise the system voltage, to state 48V. It will, in any case, give indistinguishable watt from above (48V x 5A = 240W). Multiplying the system voltage reduces the voltage drop by 1/fourth.

While the size and length of the cables involve system plan and establishment, for the nature of cables the Ministry of New and Renewable Energy (MNRE) in India specifies that cables stick to IEC 60227/IS 694 or IEC 60502/IS 1554 (Part I and II). It pays to acclimate what these specification norms state.

- ✓ **Temperature**

Solar cells perform preferred in the cold fairly over in hot climate, and as things stand, panels are rated at 25°C which can be significantly not the same as the specific open-air circumstance. For each degree ascend in temperature above 25°C the panel yield decays by about 0.25% for indistinct cells and about 0.4-0.5% for crystalline cells. In this manner, in sweltering summer days panel temperature can without much of a stretch reach 70°C or more. What it implies is that the panels will put out up to 25% less power compared to what they are rated for at 25°C. Therefore a 100W panel will produce just 75W in May/June in many parts of India where temperatures reach 45°C and past in summer and electricity request is high.

Solar panels are tried under lab conditions, called STC (Standard Test Conditions): at an Irradiance (light) dimension of 1000W/m2 with a temperature of 25°C. Be that as it may, in reality, these conditions are always changing, so the panel yield is not the same as the lab conditions. In this way, other specifications are reported, called NOCT (Nominal Operating Cell Temperature). It is the temperature reached by open circuit cells in a module under the accompanying conditions:

Irradiance (light) falling on the solar panel at 800W/m2; Air temperature of 20°C; Wind speed at 1m/s; and the group is mounted with an open back (air can circulate behind a panel).

Most great quality panels accessible today in India have NOT estimations of 47±2°C. Lower the NOCT the better it is expected to perform in more humid climates.

The temperature coefficient of the rated watt power, Pmax, is another vital parameter.

Model: EMMVEE solar panels have NOCT of 48±2°C and temperature coefficient of rated power - 0.43% per K. Moser Baer panels have NOCT of 47±2°C and temperature coefficient of rated power - 0.43% per K for panels up to 125Wp; their higher power panels have NOCT of 45±2°C and temperature coefficient of rated power - 0.45% per K.

✓ Shading

Preferably solar panels ought to be located such that there will never be shadows on them because a shadow on even a little piece of the panel can have a shockingly large effect on the yield. The cells inside a panel are ordinarily all wired in an arrangement, and the shaded cells affect the current stream of the entire panel. Be that as it may, there can be circumstances where it cannot be maintained a strategic distance from, and therefore the effects of partial shading ought to be considered while arranging. On the off chance that the affected panel is wired in arrangement (in a string) with different panels, at that point, the yield of every one of those panels will be affected by the partial shading of one panel. In such a circumstance, a conspicuous arrangement is to abstain from wiring panels in the settlement if conceivable.

✓ Charge Controller and Solar Cell's IV Characteristics

An intrinsic normal for solar silicon cells is that the current created by a specific light dimension is steady up to a particular voltage (about 0.5V for silicon) and after that drops off suddenly. What it implies is that principally the voltage shifts with light force. A solar panel with an ostensible voltage of 12 volts would regularly have 36 cells, bringing about a consistent current up to around 18 volts. Over this voltage, the current drops off quickly, bringing about the greatest power yield being created at around 18 volts.

At the point when the panel is associated with the battery through a primary charge controller, its voltage will be pulled down to close to that of the cell. This lead to bring down watt power (watt = Amp x Volt) yield from the panel. Along these lines, the panel will have the capacity to create its greatest power when the battery voltage is close to its most extreme (wholly charged). So it structures a system in, for example, a way that the batteries ordinarily don't stay precisely full charged for long. In the midst of stormy or substantial obfuscated days, a circumstance may happen when the batteries remain in the condition of not exactly full charge. This would additionally pull down the panel voltage; consequently corrupting the yield further.

This is additionally where an MPPT (Maximum Power Point Tracking) Charge Controller comes into the picture. It takes a stab at keeping the panel at its greatest voltage and at the same time creates the voltage required by the battery. An essential charge controller averts harm of batteries by over-charging, by successfully removing the current from the solar panels (or by lessening it to a heartbeat) when the battery voltage achieves a specific dimension. Then again, a Maximum Power Point Tracker (MPPT) controller plays out an additional capacity to enhance your system efficiency.

What does the MPPT Controller Do?

Other than playing out the capacity of a fundamental controller, an MPPT controller likewise incorporates a DC to DC voltage converter, changing over the voltage of the panels to that required by the batteries, with next to no loss of power. As such, it endeavors to keep the panel voltage close to its Maximum Power Point, while providing the changing voltage prerequisites of the battery. Along these lines, it decouples the panel and battery voltages so that there can be a 24-volt system on one side of the MPPT charge controller and panels wired in arrangement to create 48 volts on the other. Along these lines, it is offering the capacity to give some charging current even in dull conditions when a primary controller would not encourage much.

Applicable gauges indicated for charge controllers and power molding units by the MNRE are IEC 60068-2 (1,2,14,30)/Equivalent BIS Std., IEC 61683/IS 61683.

✓ **Inverter Efficiency**

At the point when the solar PV system is taking into account the necessities of the AC stacks, an inverter is required. As things remain, in certifiable nothing is 100% proficient. Although inverters accompany wide running efficiencies however ordinarily moderate solar inverters are between 80% to 90% proficient.

Precedent: Su Kam's 1000 VA inverter is regularly 85% proficient; their 2KW – 5KW models have over 87% efficiency. UTL Solar Hoodie Back Up (810VA – 3000VA) models are ordinarily 80% effective and the Solar S-20 display about 85%.

✓ Battery Efficiency

At whatever point reinforcement is required batteries are required for charge stockpiling. Lead acid batteries are most ordinarily utilized. All batteries discharge not as much as what went into them; the efficiency relies upon the battery structure and nature of development; some are positively more effective than others.

The energy put in a battery amid charging Ein can be given as

$E_{in} = I_C V_C \Delta T_C$ where I_C is the steady charge current at voltage V_C for time term ΔT_C

In like manner, after it is discharged at a steady present ID, at a voltage VD amid a period ΔTD; the conveyed energy is

$E_{out} = I_D V_D \Delta T_D$

Presently composing the energy efficiency as E_{in}/E_{out} = $I_C V_C \Delta T_C / I_D V_D \Delta T_D$.

There are two kinds of efficiencies: voltage efficiency (V_D/V_C) and coulomb efficiency ($I_D \Delta T_D / I_C \Delta T_C$)

Since lead-acid batteries are typically charged at the buoy voltage of about 13.5 V and the discharge voltage is around 12 V, the voltage efficiency is about 0.88. In standard, the coulomb efficiency is about 0.92. Consequently, the net energy efficiency is around 0.80

A lead-acid battery has an efficiency of just 75-85% (this incorporates both the charging misfortune and the releasing misfortune). From zero State of Charge

(SOC) to 85% SOC the normal in general battery charging efficiency is 91%-the equalization is misfortunes amid discharge. The energy lost shows up as warmth which warms the battery. It tends to be limited by keeping the charge and discharge rates low. It helps keep the battery fresh and enhances its life.

Here we did exclude misfortunes in the electronic circuit of the battery charger which may change somewhere in the range of 60% and 80%. Along these lines, the general efficiency of the battery system can be much lower.

CHAPTER SEVEN

SOLAR SYSTEM SIZING

There are two fundamental ways that individuals approach deciding the size and specifics of the solar system they will require (solar system sizing).

1. The "Make Some Now, Add Some Later" Method

A few people handle solar PV system sizing by feeling free to fabricate a decent standard measured solar system first (like one of the systems we tell you the best way to expand on this site in Solar Panel Wiring), actualizing it onto their home and after that utilizing whatever solar power they receive in return related to control from the power organization.

These individuals can likewise add increasingly solar panels to their system later on and increment their solar power generation step by step as their assets permit. They, for the most part, fabricate less power than they will need any "learn en route" (through really utilizing it) the amount more power they will require. This solar PV system sizing technique is somewhat similar to "improvising."

After some time, they can develop their systems to give all the power they require and even, in the end, utilize no authority from their service organization by any stretch of the imagination.

This is an extremely standard methodology (to solar system sizing) for the do-it-yourselfer as it enables them to get their foot in the "solar entryway" and begin profiting from solar power, rapidly, for the least cost and without a lot of dull arranging.

2. The "Make Enough For All Your Needs Now" Method

The other way that individuals decide the measure of the solar system they will require (solar PV system sizing) is by really making sense of precisely how much power their home devours and after that building a PV system that can deal with that heap.

If it's not too much trouble take note of that regardless of whether you do choose to begin little and assemble once again time you should, in any case, make sense of what measure system your family unit will require for all your energy needs, so you have a decent broad thought of what estimate system to in the end take a stab at.

Deciding this computation expects you to do some examination in and around your very own home. All the more specifically, you should check the kilowatt utilization on your electric bill and measure the available daylight in your general vicinity.

From these computations, you can decide what number of watts the solar system you assemble should need to suit the majority of your home's energy needs.

Snap here to become familiar with the standard technique for solar system sizing - by deciding the watts your system needs to deliver to oblige all your home's energy needs (completing an energy review).

You see what number of watts, volts, and amps you'll need requirement for your appliances.

Notwithstanding whether you choose to make a PV system sufficiently enormous to oblige all or only a portion of your energy needs, you are as yet going to need to in any event see what number of watts, volts, and amps you'll be delivering and whether it will be sufficient for all (or a few) of your specific appliances and power stockpiling limit needs.

This is a compelling piece of the solar system sizing process, particularly on the off chance that you will be including solar power as you go (after some time).

You should put forth some fundamental inquiries identified with solar PV system sizing like:

What number of watts will I requirement for my specific power use?

What number of volts should my system deliver for my specific appliances?

What number of amps do I require to have the capacity to deliver solar energy quick enough for my use needs?

Question 1:

What number of watts will I requirement for my particular power use?

Watts speak to the measure of power created or utilized. Consider it similar to your "power save."

With regards to PV system sizing, you have to ensure you have enough watts to power the majority of your particular appliances.

At times the watts required for specific appliances are more than you may have straightforwardly accessible or put away. Eg. Attempting to power an icebox with a PV system that produces next to no power (watts) every hour or with a battery bank that has almost no power (watts) put away.

Expanding or diminishing the watts your system can deliver and store is cultivated by including sunlight based boards and batteries to your PV system progressively. Add more boards to make more power. Add more batteries to save more power.

So suppose you need to power a laptop computer with your close system.

You have to check your laptop's watt rating (check the sticker on the back of the computer and duplicate the volts x amps to get the watts).

On the off chance that your laptop is appraised at 72 watts, this implies it needs 72 watts of power for every hour to run. So your nearby systemgroup should likewise have the capacity to either create or give from the battery bank up to 72 watts or more for every hour to have enough squeeze to power the laptop computer.

Deciding your day by day, week after week or month to month watt utilization

So how would you decide your watt utilization for the entire month, or week or day?

The appropriate response is: You need to figure the watt-hours.

Watt-hours/Kilowatt hours

Watt/Kilowatt hours is the estimation utilized by your dynamic organization to charge you on your bill. It speaks to the number of watts devoured increased by the number of hours you expend it for. One watt-hour is equivalent to swallowing one watt of power for each one hour.

Watt-hours = # of watts devoured x # of hours.

A kilowatt is equivalent to 1000 watts. It's merely one more method for saying 1000 watts, just it seems neater and is less massive looking on your service bill. So a one-kilowatt hour is equivalent to expending 1000 watts of power for one hour.

To ascertain the measure of watts/kilowatts a particular appliance devours (and in this manner will require your nearby system group to create) you have to discover two snippets of data.

The watt rating of the appliances you will utilize.

Furthermore, to what extent you utilize every appliance.

Watts

In this way, if for the timeframe of 1 day, you utilized your 72 watt laptop for 4 hours, you would have utilized 72w x 4hrs = 288 watt-hours (that is not, in any case, a kilowatt) along these lines the measure of watts that would need to be promptly accessible from your nearby system group battery bank would be 288 watts for the entire day.

To compute the aggregate sum of watts you devour for every one of your appliances or a particular gathering of instruments, you would need to head over to those appliances, get the watt rating off each and increase each by the number of hours you would typically utilize that appliance for.

At that point include every one of the sums, and you will know around what number of watts/kilowatts of power you require your close systemsystem to have the capacity to create to oblige those appliances for the timespan that you indicate (month/week/day).

As should be obvious, you're everyday sun oriented power potential incredibly relies upon what number of watts you can catch and store amid the sunlight hours.

If your nearby system group is appraised at 300 watts complete, this implies the most your system can create/store is 300 watts of power for every hour your sun based boards are in optimal sunlight conditions, yet this number might be substantially less in non-optimal sunlight conditions.

Contingent upon the measure of your system and how long of sunlight you have accessible amid the day, you can create and store vitality in your battery bank throughout the day and use it as you require it.

With our 300-watt system model above, if you had 6 hours of optimal sunlight every day, you could store 300w x 6 = 1800 watts for every day. That is much all that anyone could need juice to power your laptop which requires 288 watts for 4 hours use (or 72 watts for every hour).

Continuously check your hardware or appliances for the correct watt rating yet to give you a thought of what's in store, here are some regular wattages for some standard appliances:

Clock radio: 10 watts

Dvd player: 40 watts

Little tv: 54 watts

Light: 60 watts

Laptop computer: 72 watts

Roof fan: 120 watts

Lcdtv: 200 watts

Hand-held blender: 350 watts

Cooler: 500 watts

Espresso producer: 800 watts

Toaster: 1000 watts

Microwave: 1000 watts

Hot plate: 1100 watts

Power saw: 1350 watts.

Vacuum cleaner: 1600 watts

Albeit a few appliances like the hot plate may appear to have a higher than ordinary watt rating contrasted with a tv, these appliances are commonly utilized for littler timeframes so the general wattage utilized adjust and isn't as large as you may think.

Essentially, the more watts your nearby system group has, the more power you can create and store in your battery bank for use at whatever point you need.

Question 2:

What number of volts should my system deliver for my particular appliances?

Volts speak to the weight of electrical flow (the push).

With regards to PV system sizing, you have to ensure you have enough volts in your system to power your particular appliances (which likewise have a volt rating on them).

On the off chance that you are operating an appliance with a high voltage rating, you will require your system/battery bank to have that equivalent voltage (entirely minimal higher) to supply enough power to it for it to work.

Expanding or diminishing the voltage is practiced through the course of action/wiring of your sunlight based boards and your battery bank.

So suppose you need to power your laptop computer with your nearby system group.

Volts

You have to check it's volt rating. This ought to be on a sticker situated on the underside of the computer itself.

If your computer is evaluated at 24 volts, your close system should likewise have the capacity to deliver up to 24 volts or more to power that gadget.

It's alright to power a gadget evaluated at lower voltage with a system that produces a higher voltage, however, if you attempted it the other route around you wouldn't have adequate "push" to power the more top voltage appliances.

The distinctive voltage ratings are 12 volts, 24 volts, 48 volts, 120 volts, and 240 volts.

On the off chance that your nearby system group is appraised at 36 volts, you'll have the capacity to power appliances up to 24 volts, however not 48 volts, 120 volts or 240 volts. If your close systemsystem is appraised at 54 volts, you'll have the capacity to power appliances up to 48 volts, however not 120 volts and 240 volts. On the off chance that your nearby systemgroup is evaluated at 126 volts, you'll have the capacity to power appliances up to 120 volts, yet not 240 volts. On the off chance that your close systemsystem is appraised at 252 volts, you'll have the capacity to power appliances up to 240 volts.

Continuously check your gadgets or appliances for the correct volt rating yet to give you a thought of what's in store, here are some standard voltage ratings for some necessary instruments:

Attachment in burner component (Hot plate): 12 volts

Clock radio: 12 volts

Light: 12 or 24 volts

Laptop computer: 24 volts

Air conditioning temporary worker unit: 24 volts

LCD TV: 120 volts

Tabletop broiler: 120 volts

Water warmer: 240 volts

Dryer: 240 volts

Broiler: 240 volts

Fundamentally, the more volts your close system has, the more assortments of higher voltage appliances you can power with it - if you have the watts to supply the power.

In any case, one method for getting around this is to utilize (or purchase) appliances with low voltage ratings. Envision how much cash you would spare if you got a portion of your home's appliances at the RV and camper store. You might be shocked at what's accessible in low voltage ratings.

Question 3:

What number of amps do I require to have the capacity to deliver sun-powered vitality quick enough for my use needs?

Amps speak to the force (and sum) of current and along these lines decide the extent of the wire required.

With regards to PV system sizing, you have to ensure you have enough amps to make/store power as quick as (or quicker than) you use it. What's more, you additionally need to protect you have the correct size wire to deal with the current.

So if the close system you assemble has an aggregate of 7 amps, you would need to purchase 7 amp wire. In reality, to be erring on the side of caution, you should purchase 8 or 9 amp wire, to make sure you realize it can deal with the current.

The more amps your system has, the quicker you can make/store vitality and subsequently the more vitality you will have access to utilize - that is if you have enough batteries to store it.

When you increment the amps of your nearby system group, it resembles you're utilizing a greater wire that allows more power through on the double.

If you have enough amps just as enough batteries, you can build your generation and capacity limit, so you never come up short on sun-powered power.

Expanding or diminishing the amps is cultivated through the course of action/wiring of your sun based boards. You'll additionally require more batteries to store the additional power.

Amps

So suppose you required your nearby system group to deliver power excessively quick since you had high vitality needs and you needed to exploit your large battery bank's huge limit by energizing it back quicker after use.

To do this, you would need to mastermind the boards in your close system to build the all-out amps.

Amp Hours

Amp hours speak to how much electrical flow can flow every hour, so on the off chance that your battery says 105 amp hours on it, this implies you could charge it to deliver 105 absolute amps throughout an hour.

This is intended to give you a sign of the capacity limit of the battery and to what extent it takes a battery to release.

The more amp hours you have in your battery bank, the more drawn out your absolute power hold would take to exhaust.

Since you realize how nearby system group was sizing functions, you will have a superior thought of what estimate sun based vitality system you will require - to deal with your particular appliances and family unit loads.

In the following area, we demonstrate to you how you can add many batteries to your system and increment your vitality save whenever later on. Be that as it may, on the off chance that you need to get familiar with what sorts of sun based batteries are accessible and how to pick the correct ones, click here first.

You will likewise learn (in the following segment) how to include various sun powered boards together, how to wire your boards and battery bank to deliver diverse wanted volt and amp results and how to adjust everything together so you can take advantage of your nearby systemgroup.

CONCLUSION

Sun oriented power is a monstrous wellspring of straightforwardly useable vitality and at last makes other vitality assets: biomass, wind, hydropower and wave vitality.

The more significant part of the Earth's surface gets adequate sun-powered vitality to allow low-review warming of water and structures, even though there are huge varieties with scope and season. At low magnitudes, necessary mirror gadgets can focus sun oriented vitality adequately to cook and notwithstanding to drive steam turbines.

The vitality of light moves electrons in some semiconducting materials. This photovoltaic impact can do great scale power age. Be that as it may, the present low proficiency of sun based PV cells request vast regions to supply power requests.

Coordinate utilization of sun based vitality is the main sustainable methods prepared to do at last overriding current worldwide vitality supply from non-inexhaustible sources yet to the detriment of a land region of in any event a large portion of a million km2.

www.ingramcontent.com/pod-product-compliance
Lightning Source LLC
Chambersburg PA
CBHW051318220526
45468CB00004B/1397